STEM Careers

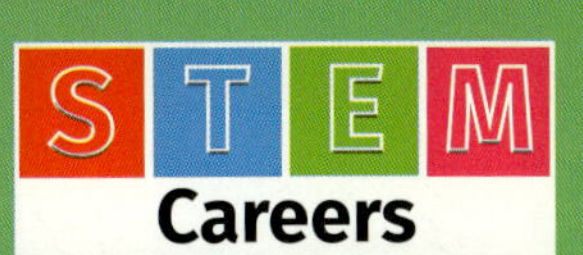

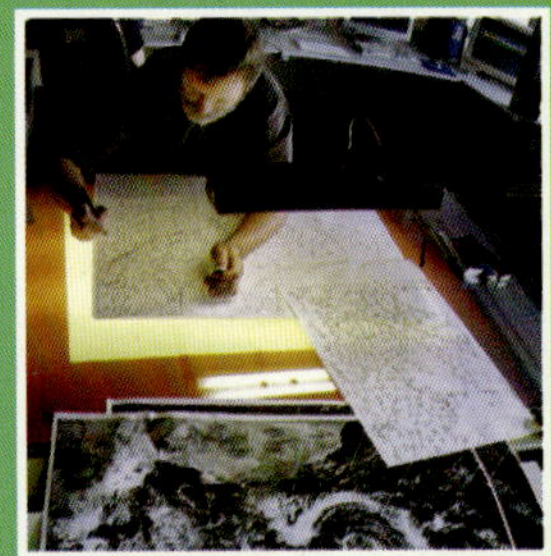

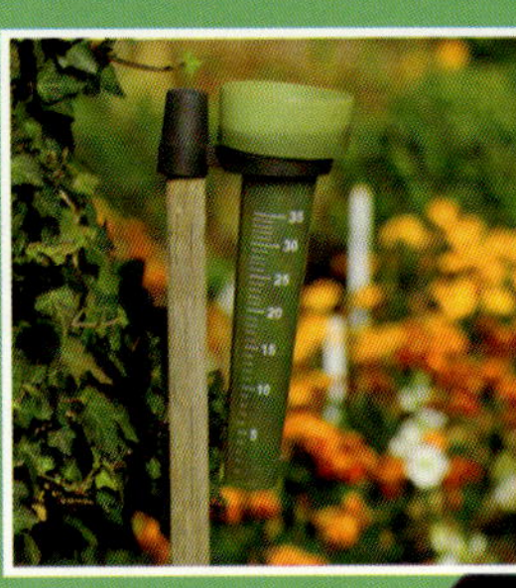

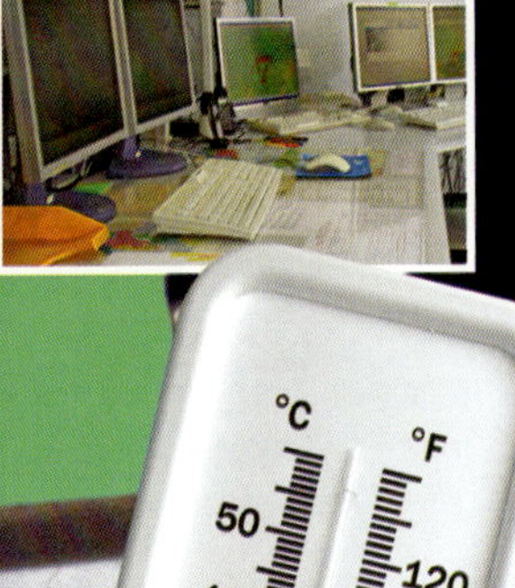

METEOROLOGIST

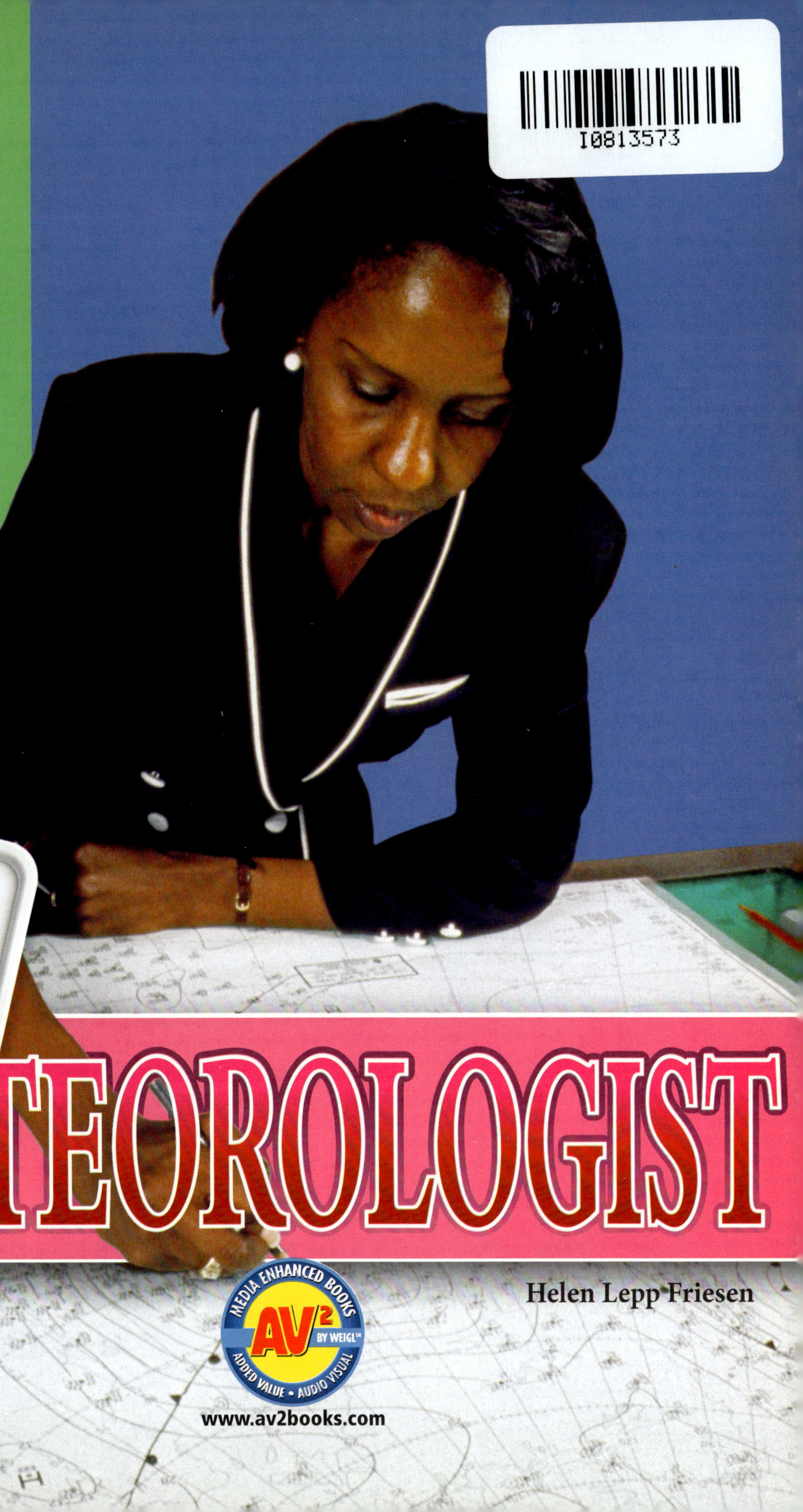

Helen Lepp Friesen

MEDIA ENHANCED BOOKS AV2 BY WEIGL ADDED VALUE • AUDIO VISUAL
www.av2books.com

Go to **www.av2books.com,** and enter this book's unique code.

BOOK CODE

AVJ66224

AV² by Weigl brings you media enhanced books that support active learning.

AV² provides enriched content that supplements and complements this book. Weigl's AV² books strive to create inspired learning and engage young minds in a total learning experience.

Your AV² Media Enhanced books come alive with...

Audio
Listen to sections of the book read aloud.

Video
Watch informative video clips.

Embedded Weblinks
Gain additional information for research.

Try This!
Complete activities and hands-on experiments.

Key Words
Study vocabulary, and complete a matching word activity.

Quizzes
Test your knowledge.

Slideshow
View images and captions, and prepare a presentation.

... and much, much more!

Published by AV² by Weigl
350 5th Avenue, 59th Floor
New York, NY 10118
Website: www.av2books.com

Library of Congress Control Number: 2019938824

ISBN 978-1-7911-0926-4 (hardcover)
ISBN 978-1-7911-0927-1 (softcover)
ISBN 978-1-7911-0928-8 (multi-user eBook)

Printed in Guangzhou, China
1 2 3 4 5 6 7 8 9 0 23 22 21 20 19

052019
103118

Project Coordinators: Heather Kissock and John Willis
Designer: Ana María Vidal

Weigl acknowledges Alamy, Getty Images, and Shutterstock as its primary image suppliers for this title.

METEOROLOGIST

Contents

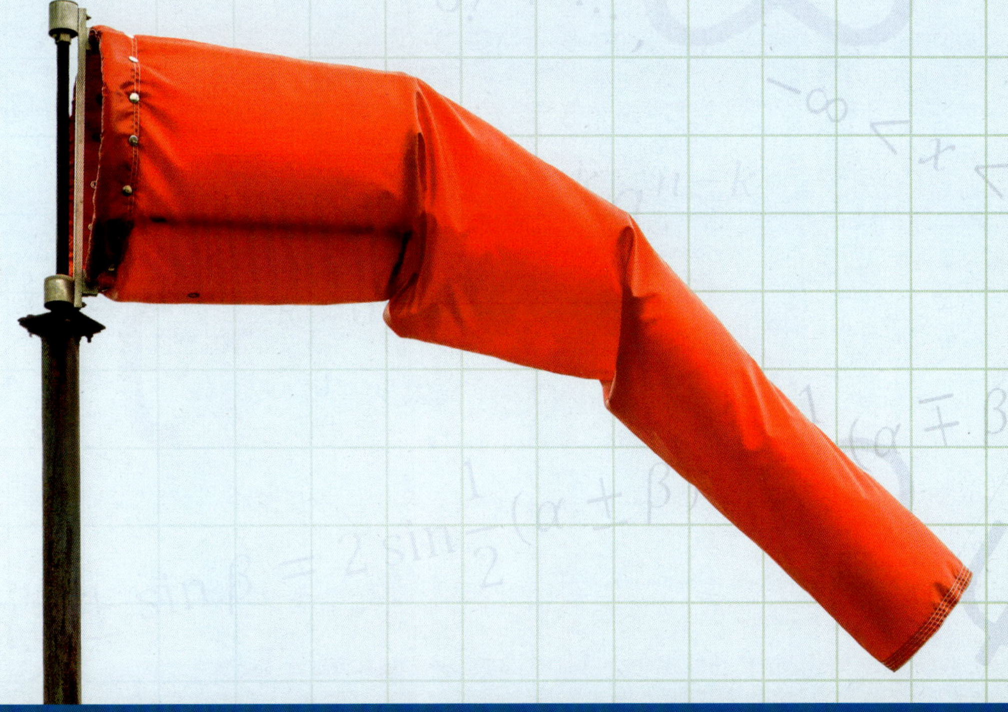

What Is a Meteorologist?

A meteorologist is a scientist who studies weather. Weather describes the **atmosphere**. Sunshine and rain are part of weather. Air temperature is an important part, too.

Meteorologists observe the atmosphere. Strong winds and rain might follow a pattern around a city. A meteorologist studies the pattern. Then, they use the information to **predict** the weather.

The work of meteorologists is important. People listen to their **forecasts**. Weather reports help people plan their days. These scientists also monitor storms. Some storms can be dangerous. This includes thunderstorms and tornadoes. When a storm is coming, meteorologists warn people. Early warning helps people stay safe. Some meteorologists study long-term weather patterns. They make predictions far into the future. Their work helps people, such as farmers, make plans.

Greek philosopher **Aristotle** wrote a book about weather in 350 BC.

Robert Fitzroy created the first weather forecast in 1861. It helped sailors avoid bad weather at sea.

Clint Youle gave the first televised weather forecast on television in the United States in 1949.

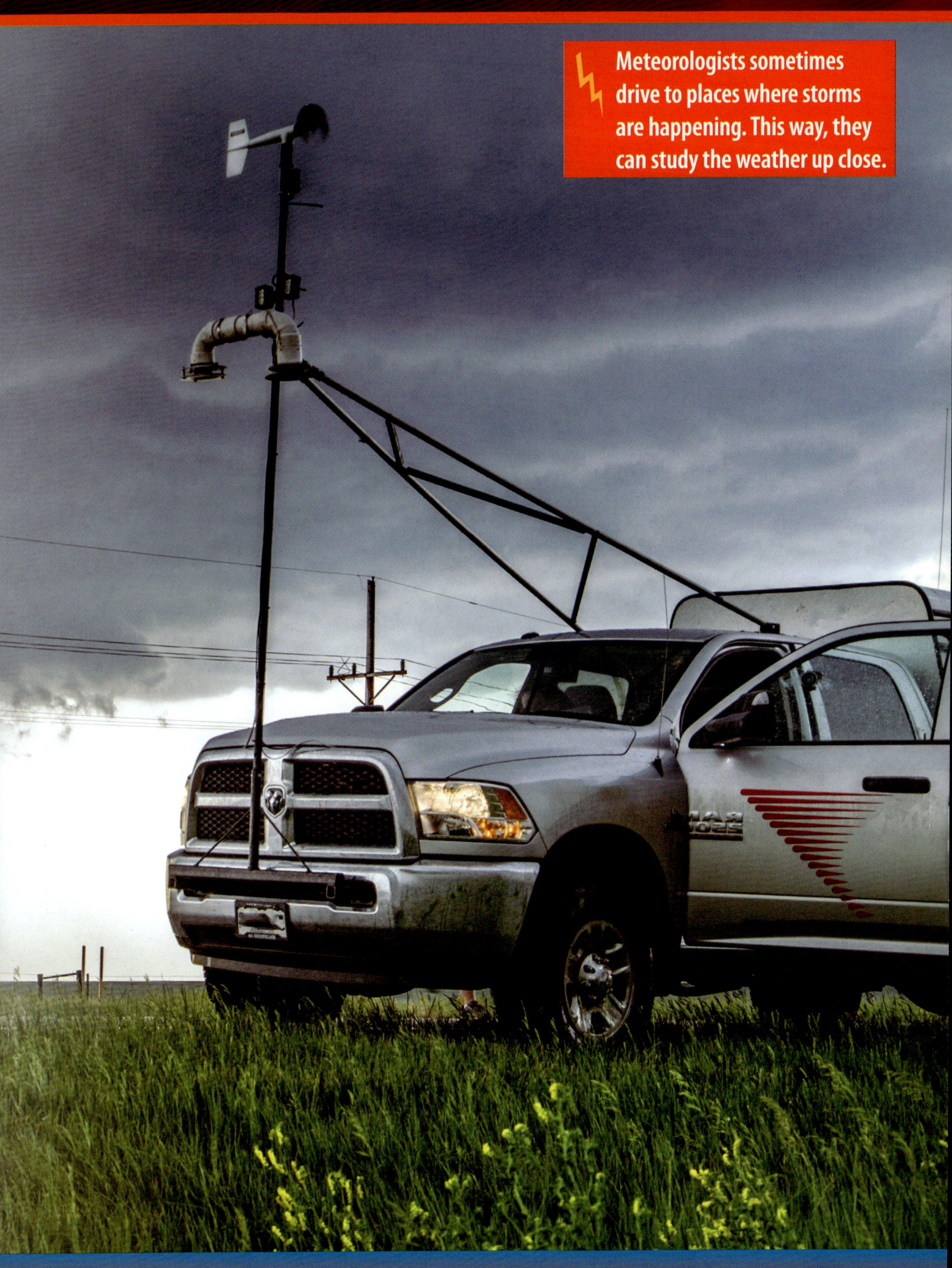

Meteorologists sometimes drive to places where storms are happening. This way, they can study the weather up close.

A STEM Job

Meteorologists work in the field of STEM. STEM stands for Science, Technology, Engineering, and Mathematics, or math. STEM is a developing field. Jobs in STEM change as new tools and techniques are created.

Meteorologists use science to do their jobs. Scientists do **experiments**. These experiments help people learn how the weather works. Meteorologists also use technology. This includes computers and **satellites**. They may work on making new technology for studying weather.

Meteorology Jobs

Meteorologists can choose from many different workplaces. Most work for private companies and government agencies.

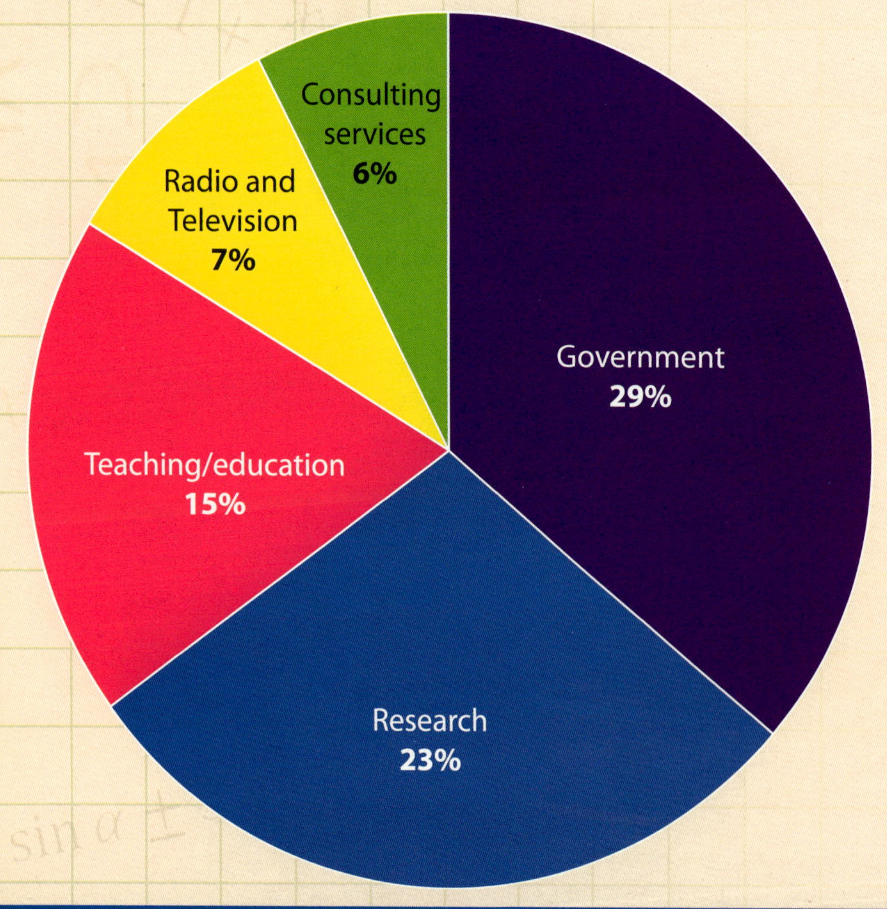

Engineers solve problems. They may work with meteorologists to build devices to measure the weather. Meteorologists with math jobs explore patterns, shapes, and numbers related to the weather. They collect **data**. The data is put into math programs. These programs make weather predictions.

There are many kinds of work in meteorology. A person who studies **climate change** is called a climatologist. A research meteorologist may do research for the government or colleges. Forecasters use advanced math and computer programs. They use information to make guesses about future weather patterns.

Satellites circle the globe. They take pictures of the weather. Meteorologists use the pictures to make forecasts.

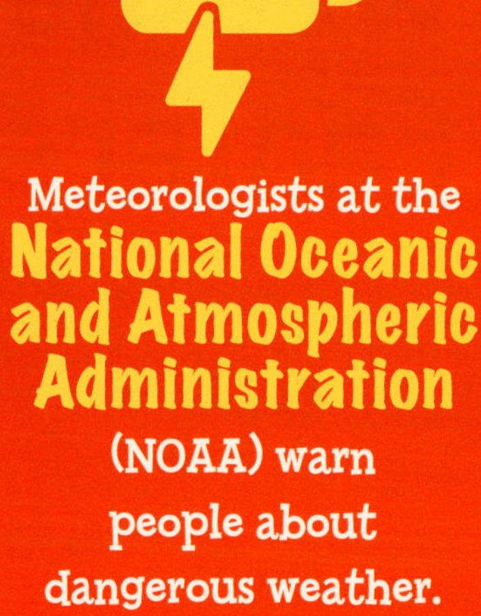

Meteorologists at the **National Oceanic and Atmospheric Administration** (NOAA) warn people about dangerous weather.

NOAA meteorologists run **three** different types of weather satellites.

There are more than **900** NOAA weather stations. They measure weather conditions up to 288 times a day.

About **2,600** meteorologists work for the United States National Weather Service (NWS).

Weather Prediction Centers

Some meteorologists work for television and radio stations. These scientists study local weather. They let people know what the weather will be. Most meteorologists in the United States work at government sites. They help leaders make important decisions based on weather. Others conduct research. They study climate and future weather.

Weather Prediction in the United States

Meteorologists work in many locations to predict the weather around the United States. They may help people know which clothes to wear, or help them keep safe when dangerous storms form.

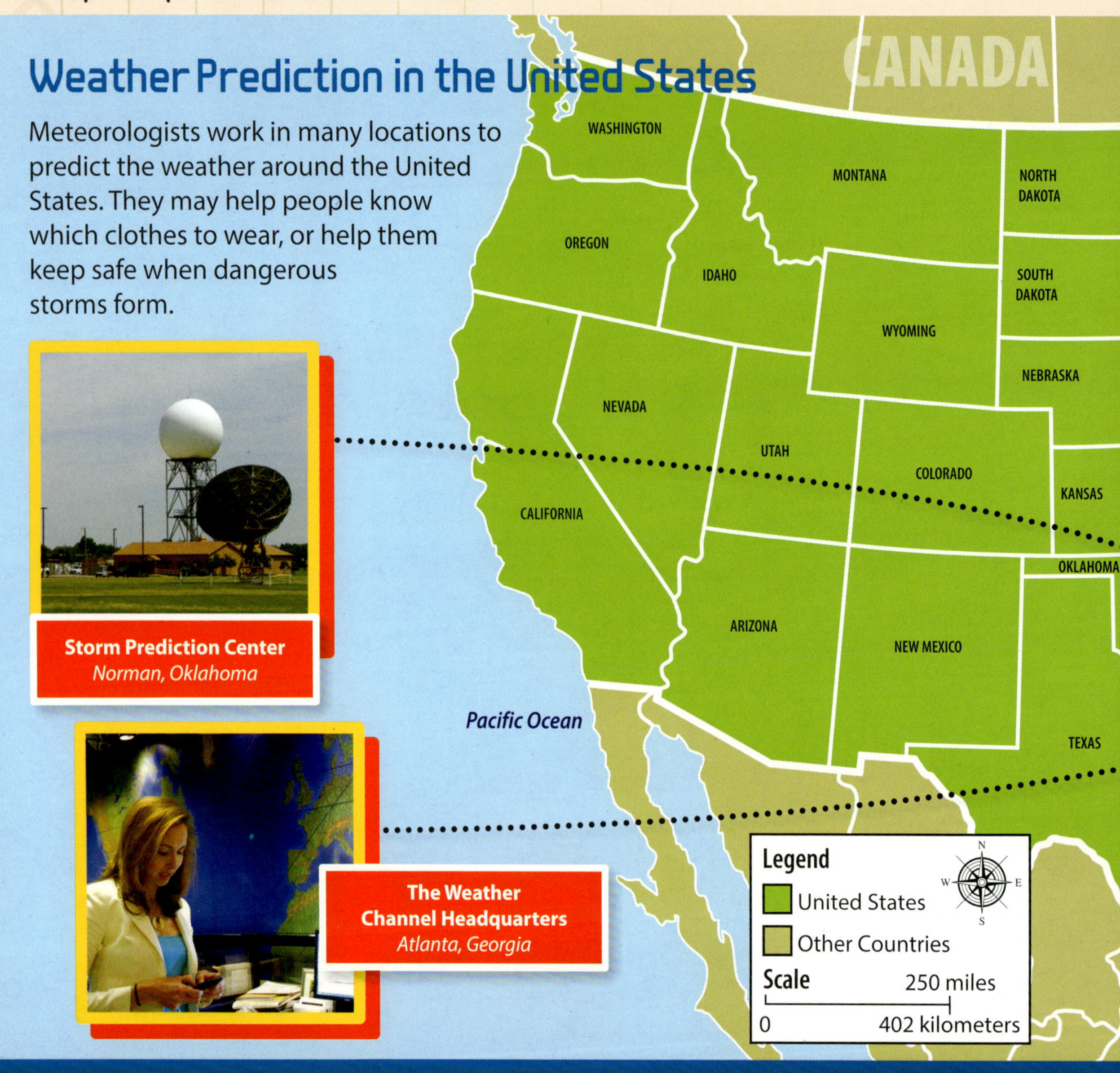

Storm Prediction Center
Norman, Oklahoma

The Weather Channel Headquarters
Atlanta, Georgia

Some meteorologists work in offices or laboratories. Often, they travel to different places to observe the weather. They collect data about air temperature and wind speed. Sometimes, they travel to see storms in person. They may observe weather from the ground or from the air. Meteorologists also visit places after a storm has hit. They look at the damage the storm caused.

A meteorologist may travel to meet with other people, including local leaders, such as mayors. Together, they talk about how they can help people prepare for weather. They work on plans to prevent damage from weather events.

AccuWeather Headquarters
State College, Pennsylvania

The National Weather Service (NWS) Headquarters
Silver Spring, Maryland

Meteorologists must be willing to report the weather as it is happening, even if it means they might get cold or wet.

All in a Day's Work

Most meteorologists use computers every day. They write forecasts and briefings. Briefings explain the current weather conditions. Briefings are sent to different groups. This includes businesses and government offices. These groups then plan the day's work and events. Meteorologists also write reports that cover long-term weather predictions. Reports require research. Often, meteorologists work in teams to write reports.

Many meteorologists work extra hours during dangerous weather. They come to work early and leave late. Sometimes, they miss meals and get less sleep. These meteorologists are on the news. They share weather warnings. This includes harmful **ultraviolet (UV)** levels. People can then protect their skin and eyes. Meteorologists also track air quality. They alert people when the air might be unhealthy.

Sample Daily Schedule

On average, one or two major hurricanes hit the United States each year. Meteorologists are busy during these dangerous storms.

2:30 AM

The meteorologist arrives at the National Hurricane Center in Miami, Florida. He studies the most recent hurricane data. The data comes from satellites, aircraft, ships, and land-based **radar**.

3:00 AM

The meteorologist looks at the hurricane's path. He contacts local weather centers in the path of the hurricane.

5:00 AM

He meets with other meteorologists. They look at storm data. The meteorologists try to figure out if the hurricane might cause tornadoes and coastal flooding.

7:00 AM

The meteorologist talks to a news reporter about the hurricane. The interview is shown on national news. He tells people how they can prepare.

8:00 AM

The meteorologist studies any changes in the hurricane's size and wind speed. The meteorologist writes a new **Tropical Cyclone Public Advisory**. It will be shown on news programs in places affected by the hurricane.

12:00 PM

The meteorologist watches the storm as it gets closer to land. He sends out **evacuation** notices to all people in the storm's path. These let people know it is time to leave and go to a safer place.

Using the Scientific Method

Scientists use the scientific method to solve problems. The scientific method is a series of steps. It begins with a question scientists want to answer. First, they find out if another scientist has asked the same question. If so, they review the scientist's results and gather information for their work. Next, they make a hypothesis, or guess about the answer to the question. Then, scientists test the hypothesis with an experiment. After the experiment, they form a conclusion. This means they study the results and decide if the question was answered. Finally, scientists share the results with the public and other scientists.

1. Ask a Question

What happens when water inside a bottle spins quickly in one direction?

2. Construct a Hypothesis

The spinning motion will create an inward force. It will form a mini tornado in the bottle.

3. Test with an Experiment

Fill a plastic bottle about ¾ full of water. Close the bottle with a cap. Turn the bottle upside down. Hold the bottle neck. Spin the bottle in a circle. Stop spinning. Look in the bottle to see what happens.

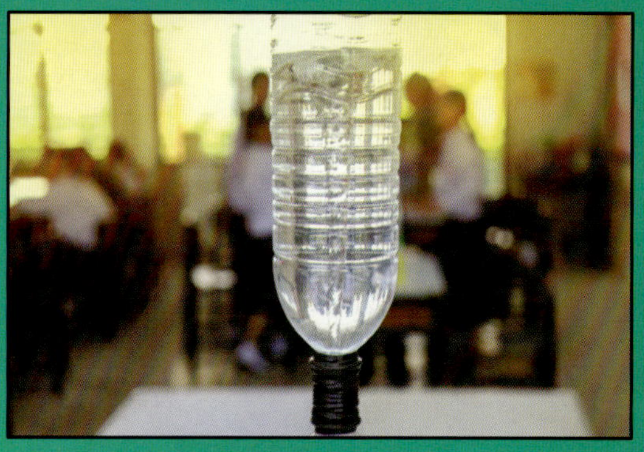

4. Did the Test Work?

The water in the bottle looks like a tornado.

5. Draw Conclusions

Spinning the bottle in a circle makes a **vortex**. That is how a tornado forms in nature.

Tools of the Trade

Most people learn about weather by seeing what is going on around them. Forecasting the weather requires more advanced tools. Meteorologists use tools and advanced technology to help in their work.

Thermometer

Thermometers measure air temperature. A thermometer is a glass tube filled with liquid, usually **mercury**. When the air around the tube is hot, the liquid expands and moves up the tube. A scale on the tube shows the temperature.

Barometer

A barometer is a device that measures air pressure. There is a metal box inside a barometer. When the air pressure rises, it pushes on the box. When pressure falls, it stops pushing on the box. Falling air pressure means weather will be stormy and wet. Rising air pressure means dry and sunny weather.

Rain Gauge

A rain gauge is a tube-shaped instrument. Some look like glass cups or buckets. They have a scale on the side to show how much rain is inside. Meteorologists use a rain gauge to measure rainfall.

Anemometers

An anemometer measures how fast the wind is blowing. It shows the wind's direction. There are cups or blades attached to a bar. The wind makes them spin. A dial attached to the anemometer shows the wind's speed and direction.

Then and Now

Technology has changed how people predict weather. People once went outside to check the temperature. They looked at the sky to see if a storm was coming. Later, meteorologists used basic tools to observe and predict weather.

Then

Some of the first radar systems were used to track ships and planes. The United States started using radar for weather in the 1940s. The radars could **detect** rain clouds. Several hurricanes surprised the United States in the 1950s. Meteorologists started using radar to track them. Soon, the technology improved. Radars could predict weather a few days in advance.

Now

Today, meteorologists use Doppler radars. These advanced radars let meteorologists see storm clouds. Scientists can also see how fast the clouds move. There are many Doppler radars across the United States. They help meteorologists predict weather up to two weeks in advance. People can even see Doppler radar images on their smartphones.

The Meteorologist's Role

Many meteorologists work for airlines. They tell pilots what to expect. Other meteorologists work for electric companies. Their work predicts energy needs during heat waves. Meteorologists also work in emergency management. These scientists often work for the government. Their predictions determine where help is sent in weather emergencies.

One way they do this is through a program called Disaster Risk Reduction. It gives people early warnings of natural disasters. People learn when they should leave their homes before a storm. They can find out about safe places to go. They can also get their homes ready. People can place sandbags around their homes. Sandbags keep water from coming inside their houses. People can also put **plywood** over their windows. This keeps their windows from breaking because of high winds. The Disaster Risk Reduction program also helps people prepare for **severe** winter weather, earthquakes, and drought.

Sandbags are filled with sand or soil. When stacked, they block the flow of flood waters.

The Difference Between Weather and Climate

Weather and climate are not the same. Climate is the weather pattern of a place over time. Weather is the state of the atmosphere at a specific time.

From 1970 to 2012, weather, climate, and water-related events caused **$2.4 trillion** in damage worldwide.

About every **40,000 years**, the tilt of Earth's axis shifts. This can cause climate change.

The average daily temperature in Quito, Ecuador is around **67 degrees** Fahrenheit (19 degrees Celsius) every day of the year.

Becoming a Meteorologist

People who want to become meteorologists like working with numbers. They enjoy doing science experiments. They are interested in nature. Would-be meteorologists are curious about how weather is created. They are interested in how humans affect the climate. Students who want to become meteorologists enjoy studying data. They are skilled at drawing conclusions.

These students should take science and math classes in high school. Students learn about heat, energy, and electricity in physics. Chemistry teaches them about **elements**. There are different elements in the atmosphere. Would-be meteorologists should learn to write and speak well.

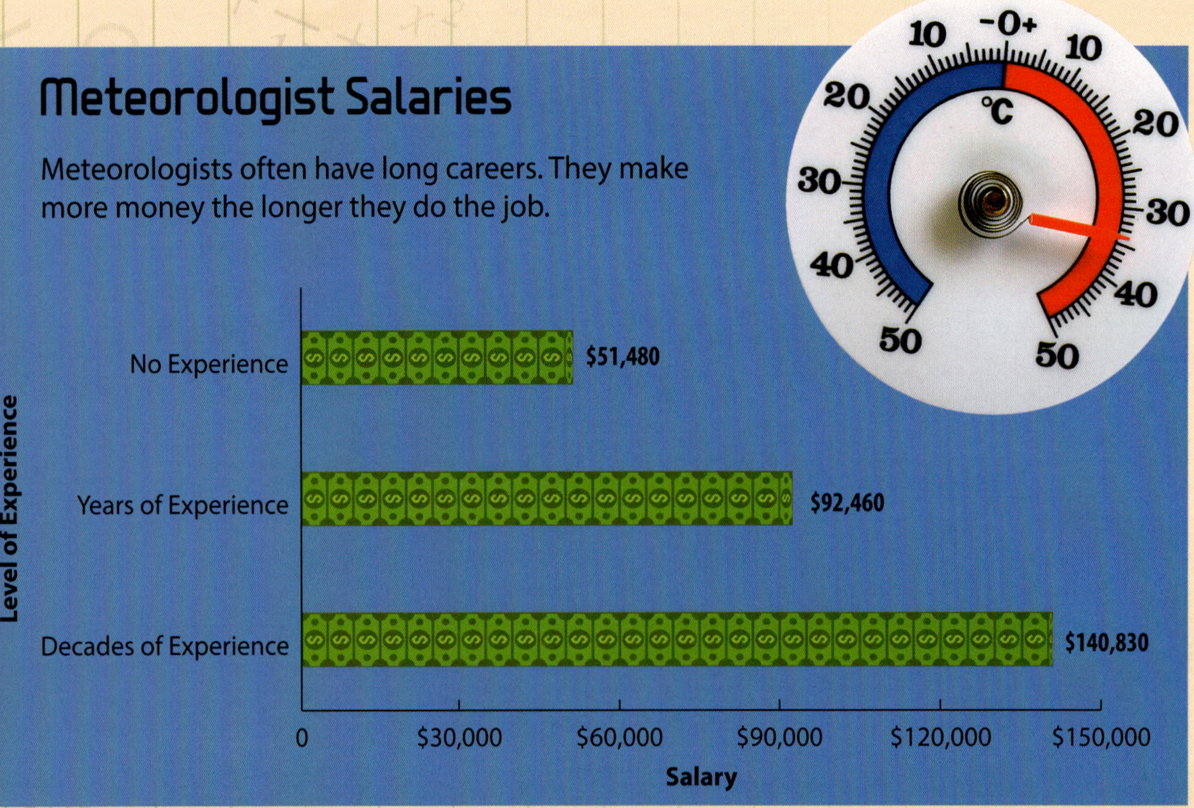

Meteorologists learn how to use many tools. Weather balloons float into the atmosphere. They send data back to meteorologists.

Is This Career For You?

Meteorology students enjoy science and data. Students should work well under pressure. They should be able to handle stressful situations. Future meteorologists need to work well with others. They should also be dependable.

Training

Training to become a meteorologist begins in college. Most of your classes will be about science and math. Outside of class, you can volunteer at a science museum or get a part-time job at a local news station. After finishing college, a meteorologist may take an **internship** to learn more about the job.

Education

Meteorologists have a bachelor's degree in meteorology. A bachelor's degree takes at least four years of college. Some meteorologists get master's degrees or PhDs. These degrees can help you get top-level jobs in meteorology. Earning these degrees takes at least four more years of college.

Application

Attend job fairs where companies are hiring meteorology students. Look for meteorologist job listings on the internet. Contact laboratories or private companies that hire meteorologists. College professors prepare students for jobs. You can talk to professors about jobs in meteorology, too.

Career Connections

Plan your meteorology career with this activity. Follow the instructions in the steps below to complete the process of becoming a meteorologist.

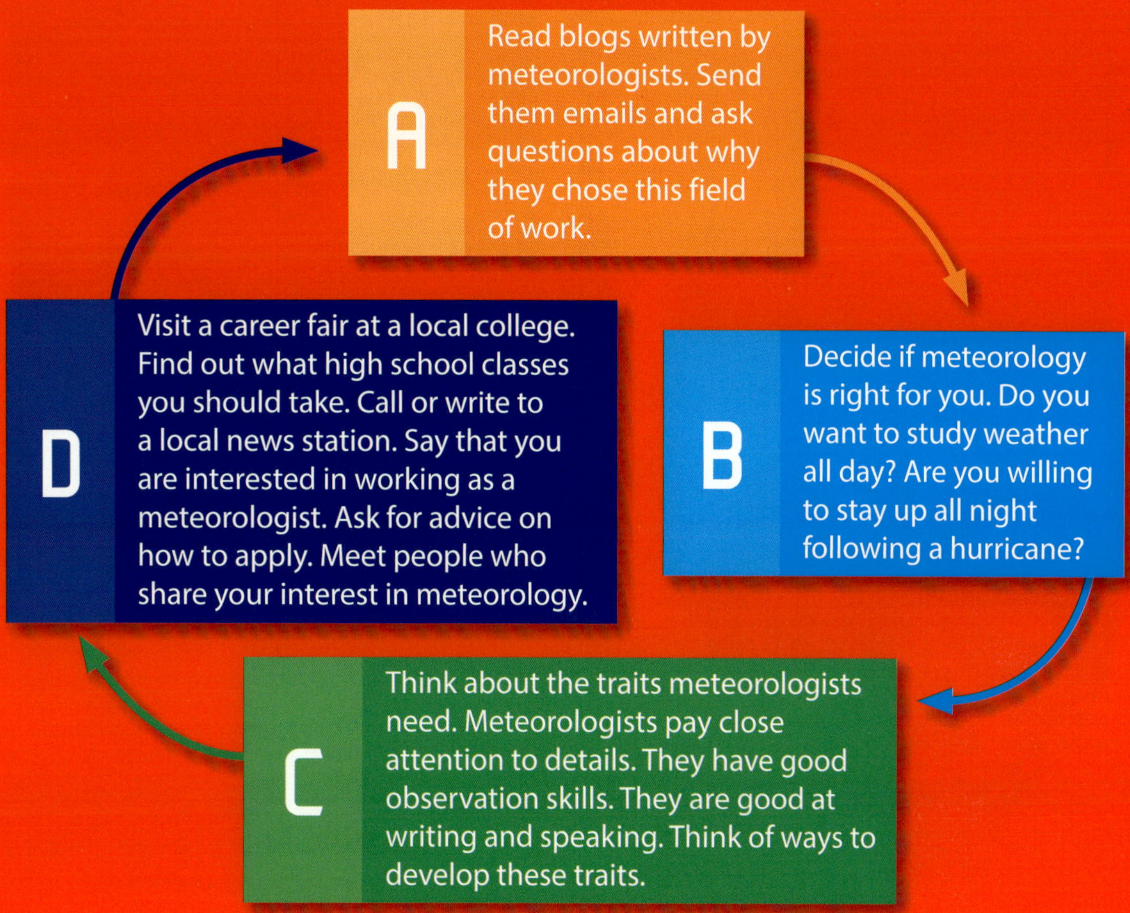

A Read blogs written by meteorologists. Send them emails and ask questions about why they chose this field of work.

B Decide if meteorology is right for you. Do you want to study weather all day? Are you willing to stay up all night following a hurricane?

C Think about the traits meteorologists need. Meteorologists pay close attention to details. They have good observation skills. They are good at writing and speaking. Think of ways to develop these traits.

D Visit a career fair at a local college. Find out what high school classes you should take. Call or write to a local news station. Say that you are interested in working as a meteorologist. Ask for advice on how to apply. Meet people who share your interest in meteorology.

1. Think about how you can get work experience while in school. Talk to your teachers. Ask them about summer jobs.

2. Research what different meteorologist jobs require. Contact employers and learn what they look for in applicants. Look for ways to develop the skills you need.

3. Follow meteorology jobs on the internet. Look for openings at colleges, news stations, government agencies, and other places that hire meteorologists.

4. Meteorology is a difficult job field. Always go to an interview prepared. Know what the job involves and how your skills match the job requirements.

Quiz

1. How many NOAA weather stations are there?

2. What does a barometer measure?

3. What is the average salary of a meteorologist with no experience?

4. What is the first step in the scientific method?

5. What is climate?

6. When did Aristotle write a book about weather?

7. How many meteorologists work for the NWS?

8. What is the job title of a person who studies climate change?

9. When was radar first used to track weather?

10. What program helps people prepare for national disasters ahead of time?

Answers

1. More than 900
2. Air pressure
3. $51,480
4. Asking a question
5. The weather pattern of a place over time
6. 350 BC
7. About 2,600
8. Climatologist
9. In the 1940s
10. Disaster Risk Reduction program

Key Words

atmosphere: the layer of air and gas that surrounds Earth

climate change: changes in Earth's weather patterns

data: facts about something, often number-related

detect: to discover the presence of something

elements: substances made from one type of atom

experiments: tests used to discover or prove something

evacuation: the removal of people from a dangerous place

forecasts: predictions about future weather

internship: a position in which a student learns how a specific job works

mercury: a heavy liquid element that changes based on temperature

plywood: a strong board made of many thin sheets of wood

predict: to tell about something that has not happened yet

radar: a device that measures a change in sound or light waves to track how severe a storm is

satellites: objects or spacecraft that orbit Earth, the Moon, or another planet

severe: causing great pain or distress

Tropical Cyclone Public Advisory: a written warning about an ocean storm or hurricane meant to help people prepare for dangerous weather conditions

ultraviolet (UV): radiation from the Sun that can cause damage to the skin and eyes

vortex: a mass of spinning fluid or air

Index

Log on to www.av2books.com

AV² by Weigl brings you media enhanced books that support active learning. Go to www.av2books.com, and enter the special code found on page 2 of this book. You will gain access to enriched and enhanced content that supplements and complements this book. Content includes video, audio, weblinks, quizzes, a slideshow, and activities.

AV² Online Navigation

Audio
Listen to sections of the book read aloud.

Book Pages
AV² pages directly correspond to pages in the book.

Video
Watch informative video clips.

Embedded Weblinks
Gain additional information for research.

Key Words
Study vocabulary, and complete a matching word activity.

Try This!
Complete activities and hands-on experiments.

Quizzes
Test your knowledge.

Slideshow
View images and captions, and prepare a presentation.

AV² was built to bridge the gap between print and digital. We encourage you to tell us what you like and what you want to see in the future.

Sign up to be an AV² Ambassador at www.av2books.com/ambassador.

Due to the dynamic nature of the internet, some of the URLs and activities provided as part of AV² by Weigl may have changed or ceased to exist. AV² by Weigl accepts no responsibility for any such changes. All media enhanced books are regularly monitored to update addresses and sites in a timely manner. Contact AV² by Weigl at 1-866-649-3445 or av2books@weigl.com with any questions, comments, or feedback.